ABUNDANT ENERGY
The Fuel of Human Flourishing

ABUNDANT ENERGY
The Fuel of Human Flourishing

Kenneth P. Green

AEI Press

Publisher for the American Enterprise Institute

Washington, D.C.

Distributed by arrangement with the National Book Network
15200 NBN Way, Blue Ridge Summit, PA 17214
To order call toll free 1-800-462-6420 or 1-717-794-3800.

For all other inquiries please contact AEI Press, 1150 17th Street,
N.W., Washington, D.C. 20036 or call 1-800-862-5801.

Green, Kenneth Philip, 1961-
  Abundant energy : the fuel of human flourishing /
  Kenneth P. Green.
     p. cm. — (Common sense concepts:ideas for a free & gen)
  Includes bibliographical references.
  ISBN-13: 978-0-8447-7204-2 (pbk.)
  ISBN-10: 0-8447-7204-6 (paper)
  ISBN-13: 978-0-8447-7205-9 (ebook)
  1. Energy policy—United States. 2. Energy—United States.
  I. Title.
  HD9502.U52.G719 2011
  333.790973—dc23

CONTENTS

LIST OF ILLUSTRATIONS   vii

ROADMAP   1
INTRODUCTION   9

1. *HOMO SAPIENS* OR *HOMO IGNIFERENS*?   17
    Fire and Human Development   19
    Our Dependence on Fire   23
    Study Questions   25

2. ENERGY AFFORDABILITY   27
    Direct Costs   28
    Indirect Costs   28
    The Impact of Indirect Costs on the Poor   36
    Study Questions   38

3. ENERGY RELIABILITY   39
    Energy Reliability and Its Impact on Related Systems   41
    The Economic Costs of Blackouts   42
    The Capacity Factor   44
    Study Questions   47

4. ENERGY AND THE ENVIRONMENT   49
    The Environmental Kuznets Curve   50
    Air Pollution and the Kuznets Curve   53

The Importance of Understanding the Kuznets Curve 56
The Deepwater Horizon Oil Spill 57
Study Questions 62

5.   ENERGY SYSTEM INERTIA AND MOMENTUM 63
Inertia and Momentum: An Overview 64
Technological Momentum 65
Labor-Pool Momentum 67
Economic Momentum 68
Study Questions 70

6.   ENERGY INDEPENDENCE AND SECURITY 71
Defining "Independence" 72
Independence and Trade-Offs 73
Study Questions 77

7.   THE DANGER OF UNINTENDED
CONSEQUENCES: THE ETHANOL FIASCO 79
Government and Unintended Consequences 80
The Case of Ethanol 81
Study Questions 95

CONCLUSION 97
ONLINE RESOURCES 99
ENDNOTES 103
ABOUT THE AUTHOR 109

LIST OF ILLUSTRATIONS

TABLES

1	Indirect Energy Consumption in the United States	30
2	Ratio of Indirect Energy Expenditures by Income	37
3	Air Pollution Trends, 1980 versus 2008 and 1990 versus 2008	54
4	Total Crude Oil and Petroleum Product Imports by Country of Origin, June 2010	75
5	Total Crude Oil and Petroleum Product Exports by Country of Origin, June 2010	76

FIGURES

1	The Environmental Kuznets Curve	51
2	Tanker Oil Spills versus Offshore-Drilling Spills, 1957–2010	59

ROADMAP

The goal of this book is to give readers the intellectual touchstones that will allow them to understand energy policy in a holistic and rigorous fashion.

In the introduction, we will discuss the subject of energy writ large, consider its role in our society, and examine the importance of understanding both energy and how energy policy might affect our society as well as others around the world.

In chapter 1, we will discuss humanity's intimate relationship to the use of energy. We will see that, rather than being addicted to or dependent on energy use, human beings have adapted, over the millennia, to ever greater energy use and that adaptation has brought with it longer lives, better health, greater wealth, and vastly expanded opportunities for self-realization and development. Neither human beings nor our technological civilization can survive without energy. Expanding access to energy, especially in the developing world, should be a high priority for those concerned with reducing global poverty and bringing hope to a desperately poor swath of humanity.

In chapter 2, we will talk about the importance of energy affordability, an issue that is of particular importance to those concerned with alleviating poverty. We will see how energy infuses virtually everything that people make or do, from raising food to providing medical services. And we will see how raising the costs of energy, as it moves into and along the chain of production of goods and services, not only raises the costs of such goods and services but also disproportionally harms the poor, who use a greater share of their income paying for energy both directly and as a component of the goods and services they consume.

In chapter 3, we will discuss the devastating consequences that unreliability in energy systems—in this case electricity—can wreak upon our energy-based society, as we study the great blackout of 2003. We will learn how unreliability can cause significant economic harms, as well as great human suffering. As the world turns to ever greater adoption of intermittent forms of energy, such as wind and solar power, it is vital that we ensure a reliable flow of electricity. Currently, technologies that can store wind and solar energy in order to increase their reliability are costly and insufficient to allow for rapid, large-scale deployment of these technologies without risking destabilization of our electricity supply.

In chapter 4, we will discuss the relationship between energy and the environment. No one can deny that energy production, distribution, and use inflict significant health and environmental impacts both locally and globally. But understanding the relationship between energy and the environment requires an understanding of the relationship between energy-stimulated societal wealth creation and environmental protection. As we will see, mainstream environmentalism has misunderstood the nature of environmental progress in general, as well as how it pertains to energy use. Contrary to the free-lunch claims of many environmentalists, we will show that environmental protection is something that can be provided only by countries that are wealthy enough to afford it. We will also show that energy use, as a key factor in production, is in fact the resource that generates that wealth. We conclude that more, and more-affordable, energy is needed to enable more environmental protection in the developed world, but most especially in the developing world, where energy poverty results in vastly higher levels of environmental despoliation and poses harms to human health.

In chapter 5, we will turn to the question of energy transitions. For decades, politicians of both parties have pledged to change our energy systems quickly, whether toward reducing imports, changing the way we distribute our energy, or, as it is in much discussion today, changing the way we produce energy, from fossil fuels to other sources such as wind power, solar power, and biofuels. What we will see, however, is that contrary to political wishes, energy systems are slow to change. Power plants, energy infrastructure, and energy-consuming equipment such as boilers, diesel trucks, airplanes, and smelters are long-term investments, with cost-recovery cycles that span decades. Like an aircraft carrier, our energy system has both inertia and momentum that makes it slow to accelerate, slow to decelerate, and slow to change course. Politicians specialize in putting forward "aspirational" goals for rapid, massive changes to energy systems. But to the extent we divert resources toward such goals, we are fighting against the tide of energy-system history, and we risk neglecting existing energy systems that we will depend on for decades to come.

In chapter 6, we will examine the slippery concept of "energy security," which has been the stated goal of many a presidency and which is a ringing call that regularly falls from the tongues of those who wish to reduce energy imports, stop enriching our enemies, prevent economic harms from energy price shocks or supply disruptions, and so on. Alas, as with most questions relating to energy, digging into the question of energy security raises more questions than it answers. Do we want to end imports from say, Canada or Mexico, two of our largest suppliers? Are we willing to see the losses in trade that will come from other countries countering any import restrictions we may impose on the things they buy from us? Are we willing to have more of our own land area consumed with biofuel farms and accept the devastating environmental impacts we have seen from even limited replacement of gasoline with corn-based ethanol? Are we willing to export ecosystem devastation as developed countries convert rainforests to biofuel plantations to slake the energy thirst of the developed world?

Finally, in chapter 7, we will explore what may be the greatest danger of all in considering quick changes to our energy society: the risk of dramatic, unintended consequences, consequences that are more likely to happen when central planners, who bear no responsibility for the consequences of their action, are given power over large elements of our energy markets.

The example we will study is the tale of corn-based ethanol, which succeeding governments have required to be incorporated into the nation's gasoline supply. We will see how an action originally intended to benefit human health and the environment by reducing air pollution has actually led to the worsening of air pollution, water pollution, wildlife endangerment, oceanic dead zones, and more. The same cautionary tale applies to wind energy and solar power, both of which are increasingly failing tests of environmental beneficence and economic affordability.

This book is not intended to convey a detailed review of our energy use as a nation. One would be hard-pressed to do that with several shelves worth of dense information. This book would, in fact, be too short to give a detailed review of energy use in a city, suburb, single-family home, or even an individual's day-to-day life. From the time we wake until the time we sleep, our consumption of energy is virtually seamless. One could easily write a long book just detailing and contextualizing where all the energy comes from to make and operate a mere handful of devices such as a laptop computer, plasma television, wind turbine, or solar panel. I will try to give enough details and concrete examples to help the reader think about energy issues, particularly energy policy—the myriad set of rules, regulations, government interventions, and private actions that cumulatively produce, import, export, distribute, and account for the costs of energy use.

Still, the primary purpose of this book is to introduce and discuss a set of energy-policy concepts that can help people better understand the energy civilization in which we live and better understand and contextualize arguments over specific energy policies. I have written this book as a series of complementary essays that can also stand alone in order to offer greater flexibility to readers who may have an interest in only one or another element of energy policy. We will start by examining the fundamental nature of our relationship to energy use and then turn to some recently proposed changes we might make to our energy systems.

After reading this book, readers confronted with a new energy proposal should be better equipped to evaluate how that proposal relates to preserving the benefits of our energy-based civilization; energy abundance; energy affordability; energy reliability; energy independence and security; energy and the environment, and the nature of energy system transitions.

INTRODUCTION

In the midst of all the debate over fossil fuels, we seem to have forgotten this fundamental role of energy in life. We think that all we need energy for is to drive our cars, fly around the world, run our electrical gadgets. But more important is that abundant energy is necessary for our way of life, for our civilization.

cont.

If that energy were to vanish, we would find ourselves once again living at the margin, and might well see the end of many things that we don't associate with an energy supply, including democracy and the freedom and creativity that leisure makes possible.

—*Daniel B. Botkin*[1]

Even though energy is all around us, and we consume copious quantities of it in virtually every form imaginable, most people only really think about energy when one of two things happens. Either they open their mail one day and have an unwanted epiphany when they realize one of their energy bills has become uncomfortably high—for diesel fuel, electricity, natural gas, heating oil, propane, and so on. Or, they suddenly have one of their energy-systems or energy-dependent devices let them down, as, for example, when the electricity goes out; the alarm clock fails; the stove fails to light; the water heater breaks down; or the car runs out of gas or has a dead battery; or they wake up one morning with a dead Kindle, netbook, iPod, Droid, or, worse, a dead coffeemaker (something that would probably disturb many Americans most of all).

But such instances have not been all that common in the United States. For many decades, Americans have had the good fortune and innumerable economic, health, and lifestyle benefits of using highly affordable energy. While most Americans will remember periods where prices spiked, such price shocks have been relatively infrequent events, usually triggered by an outside cause, such as instability in the Middle East or unexpectedly rapid economic growth in China.

This low-cost-energy blessing has not been an accident: unlike the many other countries, U.S. taxation on energy has been reasonably low, regulations have been significant but accommodated by continued access to abundant and affordable energy, and we have benefited from a highly efficient private energy sector to discover, produce, and bring energy (both in liquid form and in the form of electricity) to meet consumer demand.

Despite periodic (and highly unpleasant) power outages or equipment breakdown, Americans are generally blessed with pretty reliable energy systems. Yes, there are periodic weather-induced blackouts, and sometimes wires short out, appliances break, and the cable goes out. And, of course, people will always lose transformers and forget to plug things in. But still, most of the time, when you reach for an energy-dependent device (which you do far more than you realize, as will be discussed later), the energy is there for you on demand, 24/7, to cook your food, light and heat your home, bring your entertainment and important information to a monitor near you, and take you where you want to go in a speedy, comfortable, and generally safe manner. That's not the case for much of the rest of the world. According to *Scientific American*, "An estimated 79 percent of the people in the Third World—the 50 poorest nations—have no access to electricity, despite decades of international

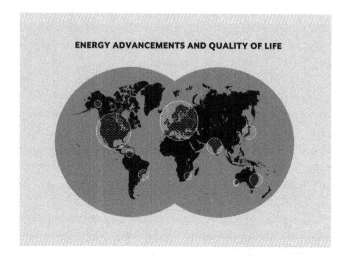

ENERGY ADVANCEMENTS AND QUALITY OF LIFE

development work. The total number of individuals without electric power is put at about 1.5 billion, or a quarter of the world's population, concentrated mostly in Africa and southern Asia."[2] And the situation is particularly acute in sub-Saharan Africa, where "several entire nations there [are] effectively non-electrified. In eleven countries, all in Africa, more than 90 percent of people go without electricity. In six of these—Burundi, Chad, Central African Republic, Liberia, Rwanda and Sierra Leone—3 to 5 percent of people can readily obtain electric power."[3] Given that we know how to produce and distribute electricity efficiently, such energy poverty is a completely unnecessary affliction in the developing world.

Americans are also blessed to have increasingly safe energy supplies, which put out ever-decreasing quantities of hazardous air and water pollutants and which operate with greater physical safety than ever before. Most people in developed countries such as the United States face much fewer health risks from the production, distribution, and use of energy than they have in the past. The same is true for most (though certainly not all) animals and ecosystems. That's not to say that energy is entirely safe (few things in life are), and energy production, distribution, and associated air and water emissions are clearly not environmentally benign—but virtually all of the energy-related trends involving human health and the environment are positive in developed countries such as the United States.

Of course, saying this will raise an immediate question from many readers: "What about climate change and greenhouse gases?" And that's an excellent question, one that would require a good 100,000 pages to discuss in any kind of depth. Without dismissing the importance of the issue—how climate-change perceptions could influence public policy is very important indeed—this book will not spend much time on the climate change issue for a very simple reason: there is really no prospect for influencing climate change through short- or even medium-term energy policy.

I happen to believe that the greenhouse effect is real—that is, that all things being equal, human greenhouse gas emissions trap a modest amount of heat in the atmosphere, and such changes may indeed pose risks to people and ecosystems. However, I also have great doubt in computerized soothsaying and do not believe that computer models can accurately capture the complexity of today's climate; the sensitivity of our climate to greenhouse gas emissions; the various feedbacks that could exacerbate or negate the effect of greenhouse gas emissions; the historical climate of the distant past; and especially the future of the climate system.

But, in a sense, one's belief in climate change is irrelevant to a discussion about energy for a simple reason: nothing that Americans (or the rest of the developed world) can do now would significantly reduce the likelihood of environmental harms, even if the people predicting disaster are correct. Yes, we could tinker around the edges with our energy system and produce some slight reductions in greenhouse gas emissions, but the reality is, China is now the world's second-largest economy, the world's largest energy consumer, and the world's largest greenhouse gas emitter. Their emission trajectory would make anything that the developed world did purely symbolic. The developing world is now behind the wheel when it comes to setting climate policy.

While happy to deploy some renewable energy production here and there, China focuses overwhelmingly on electrification using coal-fired power plants in order to grow their economy and lift their people out of energy (and other forms) of poverty. The same is true of India and other developing countries, which are focused on economic growth and lifting their people out of poverty, which is very noble goal. Even those alarmed about climate change, and those in the environmental movement admit that absent a worldwide crash-effort to reduce greenhouse gas emissions, the developed world, acting alone, could do virtually nothing to reduce actual global warming or climate change.

And, as we discuss in a chapter on energy affordability, we know that the more we push developing countries such as China and India to adopt overpriced forms of energy, the more we push them into prolonging the poverty that immiserates billions of people around the world.

Now, let's get into some of the details!

1

HOMO SAPIENS OR *HOMO IGNIFERENS?*

[Fire] provides us warmth on cold nights; it is the means by which they prepare their food, for they eat nothing raw save a few fruits.... The Andamanese believe it is the possession of fire that makes human beings what they are, and distinguishes them from animals.

—*A. R. Radcliffe*[4]

When it comes to energy, most discussions focus on the specifics: should we use less oil? Should we use less coal? More nuclear? Wind power? Solar power? Should we use less power altogether? All of these questions are important, of course, but they are too often discussed in the complete absence of context or, if you will, the big picture. And the big picture is that biology and anthropology tell us something very interesting about human beings: we are not simply beings that use energy, we are beings that exist only because we harnessed energy, and that use of energy has shaped our bodies and our cultural development for millions of years.

All known human societies, from the most advanced to the most primitive, rely on the controlled use of fire (or more advanced forms of energy) for cooking, lighting, and protection. And human beings are the only species known to do so (apes taught to smoke cigarettes don't count). Virtually all advances in human society in terms of increased security, increased food availability, increased physical comfort, increased time for study and to practice the arts, and increased ability to influence the world stem from the direct or indirect use of energy. In a real sense, while our wisdom is certainly a worthy attribute to stress, as in *homo sapiens*, we would more uniquely be identified as *homo igniferens*: Man Who Ignites Fire.

Anthropologists have long struggled to figure out exactly when primitive human beings first wove the use of fire into their lives. That turns out to be a difficult question to answer. For most of human history and prehistory, modern humans and their forebears were largely nomadic and could not build structures that were capable of surviving for hundreds of years, much less thousands or millions. And often, what did get built or occupied for a long period has been buried or erased by natural events, such as glaciations, rising sea levels, land subsidence, and so on. The small fires that early humans would likely have used would not leave much for others to find even a few years later, much less a few million. So, the further back into the past that scientists try to look, the less likely they are to find permanently occupied areas where humans might have repeatedly built fires.

Despite these limitations, archaeologists and anthropologists have discovered evidence suggesting that ancient humans were fully in control of fire, from its lighting, to its maintenance, to its use, at least one and a half million years ago.

FIRE AND HUMAN DEVELOPMENT

But some anthropologists believe that humanity's relationship with fire goes much further back in time, into our fairly distant prehistory. These anthropologists argue that primitive man's use of fire for cooking, clearing land, extending access to light, hardening tools, and protecting themselves from predators not only improved survival and reproduction but also led to changes in the human genome, brain development, digestive system, dentition, hairiness, and the many other characteristics that define what humans are today.

This tantalizing hypothesis is fleshed out in two recently published books examining the relationship of fire to human development. In Richard Wrangham's book *Catching Fire: How Cooking Made Us Human*, the focus is culinary: Wrangham's primary interest is in how cooking food increased the calories available to primitive man, changing our ancestors' lives and biology. In *Fire: The Spark That Ignited Human Evolution*, Frances D. Burton focuses on how exposure to firelight might have altered hormonal activity in the brains and bodies of humanity's primitive ancestors, leading to changes in the structure of the human brain as well as the development of more complex forms of social interaction.

Wrangham's observations about the way that fire increased the number of calories available to early humans (as well as making it possible to eat tough, stringy, large-animal game meats) suggest a human relationship with fire dating back to about two million years ago. As Wrangham observes, "Humans do not eat cooked food because we have the right kind of teeth and guts; rather, we have small teeth and short guts as a result of adapting to a cooked diet."[5] Wrangham observes that experiments with rats, snakes, and, in a few cases, human beings show that softening and cooking food makes it much easier to digest, allowing more calories and other nutrients to be extracted from the food. For example, while raw starchy foods such as potatoes are digested poorly, cooking them vastly increases their digestibility. The same is true for eggs and meat. And interestingly enough, it's not only humans that prefer cooked food; so do chimpanzees, other apes, and even insects.

Having observed that our modern-day simian relatives prefer their food cooked, and understanding that changes in diet often precipitate major changes in the biology of organisms, Wrangham casts his eye backward, looking for rapid changes in human development as proxy indicators for when fire may have been harnessed for cooking food. Wrangham ultimately settles on a major shift in human living and mobility: the change from a largely tree-dependent lifestyle to a more ground-dependent lifestyle. He notes, "Having controlled fire, a group of ancient human ancestors called habilines learned that they could sleep safely on the ground. Their new practice of cooking roots and meat meant that food obtained from trees was less important than it had been when raw food was the only option. When they no longer needed to climb trees to find food or sleep safely, natural selection rapidly favored the anatomical changes that facilitated long-distance locomotion, and led to living completely on the ground."[6] Wrangham puts this transition in lifestyle and climbing ability at the transition from *homo habilis* to *homo erectus*, suggesting that controlled use of fire was accomplished at least 1.9 million years ago.

Frances D. Burton, however, thinks that humanity's ancestral use of fire goes much further back in time, perhaps as far back as six million years. At first, Burton speculates, this association would be simply a matter of learning to recognize that fire, which is quite common in nature, has benefits, such as leaving behind a bunch of tasty cooked insects, for example. And with an estimated eight million lightning strikes hitting the earth every day, it's not unlikely that our ancestors would have seen a lot of it. Our ancestors may have noticed that smoke from a smoldering tree stump keeps away noxious flying insects, makes for warmer evenings, and so on, increasing their desire to stay around spontaneously created fires. Later, the association would have progressed to maintaining naturally kindled fires and, eventually, to discovering how fire is started and gaining full control of this most potent force of nature. Burton also observes that humans aren't the only ones to like their food cooked, and humanity's earliest attractions to fire may simply have been an observation that after a fire passes, there are a lot of tasty roasted nuts lying around, not to mention crispy fried insects.

Burton's most interesting hypothesis, however, is that harnessing fire could have influenced a hormone that is integrally involved in a vast array of biological cycles. Melatonin, the production of which is suppressed in response to light exposure, is particularly interesting to Burton because it is also involved in regulating sleep cycles, reproductive cycles, the onset of puberty (and hence reproductive age), and other fundamental traits that shaped human development. Burton sums up her view thus: "I conclude therefore that motivation, ability, circumstance, and environment merged to inaugurate this unique relationship to fire around 6 m.y.a. [million years ago]. My view of human evolution is that the acquisition of fire was the engine that propelled the incredibly fast evolution of humans. Directly or indirectly, it affected cognitive processes, social processes, genetic systems, reproduction, the immune system, and digestion, among others. It may even have enhanced hair loss."[7]

OUR DEPENDENCE ON FIRE

Whether six million years or two million years, humanity's association with fire, and thus our intricate weaving of energy into our lives, is clearly a distinguishing attribute and one that has shaped us in ways numerous, irreversible, and profound. Understanding energy starts

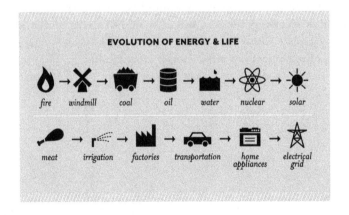

with understanding and accepting that we are no longer capable of surviving long, and certainly not very well, without access to fire in its many forms, whether it is cooking our food, heating our homes, generating our electricity, powering our cars, or making the clothing that we now need in cold weather because fire let us shed the hairy coats of our ancestors. Politicians like to talk about how Americans are "addicted to oil," or "addicted to cheap energy." It would be more accurate to say that humans are biologically and culturally adapted to reliance on energy. Are we addicted or adapted? It makes a big difference in how one perceives the role of energy in our civilization.

Having discussed the centrality of energy to the human experience, we will now take up the most important question when it comes to energy policy, that of energy affordability. For some people in the United States, and many around the world living in a state of energy poverty, no discussion of energy policy is complete without an understanding of how our energy policy choices can affect energy affordability.

STUDY QUESTIONS

1. Do you agree or disagree with the idea, explained above, that human beings have been fundamentally and irreversibly altered by their historical use of energy? Please explain your view in 400–500 words, with references to source materials supporting your view.

2. Imagine yourself lost in the wilderness for several weeks, without food or water but with the ability to make fire. Explain, in 400–500 words, how you might use that fire to ensure your survival and what specific benefits the use of fire might provide you.

2

ENERGY AFFORDABILITY

There is no substitute for
energy. The whole edifice
of modern society is built
upon it.... It is not "just
another commodity" but the
precondition of all commodities,
a basic factor equal with air,
water, and earth.

—*E. F. Schumacher, 1973*

Energy costs are experienced in many ways in American society. There is, of course, the cost of energy you pay for directly, such as your monthly electricity bill, your gas bills, your gasoline, and so on. But people also pay for energy that they consume indirectly, that is, in the goods and services they consume.

DIRECT COSTS

In 2005, the date of the last Residential Energy Consumer Survey, the average American household spent about $1,800 on nontransportation-related energy use, including electricity, natural gas, fuel oil, kerosene, and liquid petroleum gas. About $1,122 of that was spent on electricity, $471 was spent on natural gas, $115 was spent on fuel oil, and $100 was spent on fuel oil. The amount spent on kerosene was relatively trivial.[8] On top of that, the average American household in 2005 paid about $2,000 per year for gasoline.[9] Considering that the average household income was about $46,000 in 2005, direct energy expenditures would have consumed a little over 8 percent of the household budget.

INDIRECT COSTS

But that's just the beginning of America's energy bills. In addition to the direct use of energy, Americans consume a great deal of energy indirectly. Consider a simple cotton T-shirt. Energy is used to grow and harvest the cotton. More is used to transport that cotton to a factory. Still more energy is used to process the cotton, bleach,

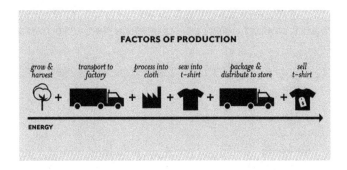

FACTORS OF PRODUCTION

grow & harvest   *transport to factory*   *process into cloth*   *sew into t-shirt*   *package & distribute to store*   *sell t-shirt*

ENERGY

and dye and to weave the cotton into cloth. More energy is used to package the T-shirt, get it to a store, and so on. And even that's only the tip of the energy iceberg, because it took energy to make the machines used throughout the process, as well as the dyes and chemicals. And we haven't even mentioned keeping the lights on at the various factories, moving the workers around, powering the looms, using the washing machines, and so on. All of the energy used at each stage of production factors into the final price you pay for whatever good or service you consume.

My colleagues and I at the American Enterprise Institute calculated just how much energy is used indirectly as a component of the various goods and services that we consume in our daily lives. What we found surprised us. For example, it turns out that nearly half (46 percent) of what people pay for energy comes "embodied" in the various goods and services that they

TABLE 1. INDIRECT ENERGY CONSUMPTION IN THE UNITED STATES

Indirect Energy Use by Category

	PERCENT OF TOTAL INDIRECT ENERGY CONSUMPTION
Health care	27.5
Food	23.7
Transportation	12.0
Housing	9.9
Entertainment	5.8
Clothing and shoes	3.7
Financials	3.4
Beauty	2.9
Religion	2.8
Education	1.5

Source: From Kenneth P. Green and Aparna Mathur, "Measuring and Reducing Americans' Indirect Energy Use" (Washington, DC: American Enterprise Institute, 2008), available at http://www.aei.org/docLib/20081204_EEON02g.pdf.

use, and nearly half of that comes down to two somewhat important things: food and health care. Transportation, another important part of our economy, comes in third. Table 1 shows the top ten ways that Americans consume that indirect energy.

Aside from the simple percentages, one can derive a lot of useful information in looking at how Americans consume their indirect energy. As mentioned earlier, health care and food are the two big-ticket items in terms of Americans' indirect energy consumption. But there's additional data within those broad categories.

Within health care consumption, for example, about half (almost 47 percent) of the indirect energy consumed involves the preparation of pharmaceuticals. Physician services come next, accounting for roughly 18.5 percent of health care's indirect energy use. Now that the United States has adopted public policy guaranteeing access to health care for everyone, it's easy to see how higher energy rates are going to influence the government's future health-care budgets.

Indirect energy in food is another interesting subject with policy relevance, and even relevance to individual consumer decisions. Many young people are interested in vegetarianism, and some vegetarians claim that being vegan is good for the earth because it uses less energy. And there is, indeed, significant variance in the energy intensiveness of the production of various kinds of meat. In general, processing larger animals is less efficient than processing smaller ones. A study conducted in Sweden found that chicken had the lowest energy consumption of the types of meat studied, at 18,500 calories per pound produced. Pork and lamb had 21,000 calories and 23,000 calories per pound, respectively, and beef had energy inputs of up to 40,000 calories per pound. This is a similar finding to other studies.

Moreover, we found that, in general, a vegetarian diet is more energy efficient than a nonvegetarian one, which is something that environmental groups regularly highlight. Overall, however, we found that meat consumption is only responsible for slightly more indirect energy use than fruit and vegetable consumption, and there are wide variations in how many calories it takes to produce different meats. This suggests that although growing vegetables in your backyard is certainly more energy efficient than buying beef, cutting out meat may not be the most cost-effective way for Americans to reduce their energy consumption.

Recently, young people have taken an interest in the idea of "local production" of food, in what some people call the "locavore" movement, which seeks to reduce energy use and associated environmental impacts by eating foods that are grown locally. This is related to the idea of indirect energy consumption, and it plays into our intuition. After all, wouldn't it be better to eat something local than to have it shipped across the country, with all the energy consumption that goes along with transportation?

Alas, as with all things involving energy, things are not as simple as they may seem. As author Stephen Budiansky observed in an article he penned for the *New York Times*, "It takes about a tablespoon of diesel fuel to move one pound of freight 3,000 miles by rail; that works out to about 100 calories of energy. If it goes by truck, it's about 300 calories, still a negligible amount in the overall picture. (For those checking the calculations at home, these are 'large calories,' or kilocalories, the units used for food value.) Overall, transportation accounts for about 14 percent of the total energy consumed by the American food system."[10]

Budiansky goes on to point out that what you do with your food once you get it home is far more important than where you get it from:

> The real energy hog, it turns out, is not industrial agriculture at all, but you and me. Home preparation and storage account for 32 percent of all energy use in our food system, the largest component by far.
>
> A single 10-mile round trip by car to the grocery store or the farmers' market will easily eat up about 14,000 calories of fossil fuel energy. Just running your refrigerator for a week consumes 9,000 calories of energy. That assumes it's one of the latest high-efficiency models; otherwise, you can double that figure. Cooking and running dishwashers, freezers, and second or third refrigerators (more than 25 percent of American households have more than one) all add major hits. Indeed, households make

up 22 percent of all the energy expenditures in the United States. Agriculture, on the other hand, accounts for just 2 percent of our nation's energy usage; that energy is mainly devoted to running farm machinery and manufacturing fertilizer.

Researcher Hiroko Shimizu derives slightly different values for the transportation component of food, but her conclusions are fundamentally the same: "Production technologies matter: 'Food miles' refer to the distance food travels from farms to retailers. In the American case, the food production stage (planting, irrigating, harvesting, using heated greenhouses, applying fertilizers and pesticides, etc.) contributes far more greenhouse gas emissions (83%) than the food miles segment (4%). Therefore, the resources needed to produce food matter a lot more than how close a production venue is to consumers. As a rule, the alleged energy savings attributable to increased local purchases is dwarfed by the additional inputs required in less productive locations."[11]

The food mile issue comes in for still more scrutiny in a longer article by Shimizu, coauthored with University of Toronto professor Pierre Desrochers. In "Yes, We Have No Bananas: A Critique of the 'Food

Miles' Perspective," the authors discuss the various studies that examine the transportation dimension of food consumption as it pertains to environmental impacts, social impacts, health impacts, and economic impacts. They observe that "another largely overlooked issue is the way consumers handle their food. Garnett (2006) points out that 25 percent of all produce grown ends up as waste. Another British study conducted by the Waste & Resources Action Programme (2008) analyzed the trash of 2,138 households and estimated that more than 6.7 million tons of food—roughly a third of the food bought by consumers—was thrown out in the United Kingdom every year. According to the report's authors, 61 percent of this food waste (consisting mostly of fresh fruits, vegetables, and salads and amounting to approximately 70kg/year/ person) could be avoided with better shopping and meal planning. Food waste costs were estimated to be on the order of £10.2 billion (about $19.5 billion USD) and the cause of 18 million tons of CO_2 emissions per year in the United Kingdom—an amount equivalent to the annual emissions of one-fifth of the British car fleet during this time period."[12]

Desrochers and Shimizu conclude that "the evidence presented suggests that food miles are, at best, a marketing fad that frequently and severely distorts the environmental impacts of agricultural production. At worst, food miles constitute a dangerous distraction from the very real and serious issues that affect energy consumption and the environmental impact of modern food production and the affordability of food."

THE IMPACT OF INDIRECT COSTS ON THE POOR

But the most important information we discovered when we studied indirect energy consumption relates to how energy prices affect different parts of the population, particularly the poor versus the better off. It has long been known that the poor spend a greater share of their income on energy than do the better off. It is not simply that the poor have less income that causes this phenomenon. An aggravating factor is that the poor often live in older, less energy-efficient houses and apartments, drive older, less energy-efficient cars, and often have to drive them longer distances to work.

We found that the same dynamic holds true for the indirect energy consumed by Americans. Table 2 shows the ratio of indirect energy expenditures to income for the population by income deciles in 2003, the most recent year for which full data were available. Notice

**TABLE 2. RATIO OF INDIRECT ENERGY EXPENDITURES
BY INCOME**

Distribution Across Income Classes, 2003

DECILE	RATIO OF INDIRECT ENERGY EXPENDITURES TO INCOME (PERCENT)
Bottom (poorest)	5.05
Second	3.72
Third	2.86
Fourth	2.46
Fifth	2.12
Sixth	1.99
Seventh	1.79
Eighth	1.65
Ninth	1.49
Top	1.33

Source: Kenneth P. Green and Aparna Mathur, "Indirect Energy and Your Wallet" (Washington, DC: American Enterprise Institute, 2008), available at http://www.aei.org/outlook/100017.

that the poorest people in American society pay the highest percentage of their income on indirect energy (5 percent), while the richest only pay about 1.3 percent. The implication of this finding is that government policies that raise the cost of energy have the greatest impact on the poor, not only directly as they gas up their car or flip on their lights but also as they consume goods and services across the economy.

We should remember, too, that affordable energy isn't only important to you as a consumer of both direct and indirect energy. As a prime input to economic productivity, energy costs affect the entire economy. From the time you awake to the time you sleep, you're consuming energy both directly and indirectly. If energy costs go up, so does the cost of everything else. When that happens, consumption declines and unemployment rises. As with many things, the poor are the most harmed by actions that undermine energy affordability in the United States, and still more are harmed in the most poverty-stricken reaches of the world.

Of course, energy affordability is not the whole picture. The production, distribution, and consumption of energy have many impacts on society and the environment. We next turn to the question of reliability, another key issue to ponder as we consider the direction of our nation's energy policy.

STUDY QUESTIONS

1. In approximately 500 words, summarize how Americans use energy, both directly and indirectly.

2. Discuss (in approximately 500 words) how higher energy prices affect Americans differently, based on their income.

3. Discuss how bringing affordable energy to people in developed countries could alleviate their most common health and nutritional problems.

3

ENERGY RELIABILITY

As dawn approached in the New York area, lights were reported flickering back on in Times Square, on Fifth Avenue, much of Staten Island, parts of Brooklyn and the Bronx, parts of Westchester County, N.Y., and parts of New Jersey and Connecticut.

cont.

But the New York metropolitan area is still in a major mess, with full power still not back, meaning that subway and train systems are also not back. Transit officials have said even when the power does come back, it will take as much as six hours for trains to start running normally.

But some folks will have to go to work anyway today. And despite today's 90 degree weather forecast, some are facing the inevitable: saddling up in the most comfortable socks and shoes they've got, heading out on the long walk to work.

—Jaime Holguin[13]

On the evening of August 14, 2003, about fifty-five million people found out just how important reliable energy is when a massive power failure shut down large parts of the midwestern and northeastern United States, as well as the Canadian province of Ontario. It was the second most widespread electricity blackout in history.

Daniel B. Botkin, author of *Powering the Future*, remembers the view from his apartment in Manhattan, an apartment that was spared by virtue of having its own generator: "The view from our lofty perch showed an island of light—Penn South, our ten-building cooperative complex—in a sea of darkness that was the rest of the city. Even the colored lights atop the Empire State Building had gone dark."[14]

The great blackout of 2003 was a classic example of the old proverb, "For want of a nail":

For want of a nail the shoe was lost.
For want of a shoe the horse was lost.
For want of a horse the rider was lost.
For want of a rider the battle was lost.
For want of a battle the kingdom was lost.
And all for the want of a horseshoe nail.

In the case of the 2003 blackout, the cascade of events started when a generator in Ohio went off-line while demand was high, placing additional strain on some rural high-power lines elsewhere in the state. Those lines heated up and sagged, dropping far enough toward the ground to bring them near some overgrown trees, which caused the lines to further overload and be automatically shut down. A cascading set of failures along the midwestern, northeastern, and Canadian eastern grids ultimately led to the shutdown of more than one hundred power plants.

ENERGY RELIABILITY AND ITS IMPACT ON RELATED SYSTEMS

We also need to understand that it wasn't just the power going off that caused problems. Without power, many water systems were adversely affected. Significant swaths of transportation infrastructure were crippled, and trains were actually "lost." Cellular phones became paperweights.

Botkin details some of the misery:

> Rush-hour commuters were stalled everywhere,
> Perhaps the worst spot to be driving was in or into
> the Detroit-Windsor Tunnel, between Michigan
> and Ontario. About 27,000 commuters used it
> daily. Some were stuck in the dark. People waited
> seven hours in the line to go through. Amtrak
> trains stopped running: the railroad was without
> electric signals, and even more surprising, no
> one had any idea where any train was. Even the
> main train from Detroit to Chicago was lost
> temporarily. You might ask why people stuck
> on trains didn't use their cell phones to give
> their trains' locations. The answer is that all cell
> phones stopped working too − the towers that
> sent and received their signals were powered
> by the grid. Even Detroit's homeland security
> director couldn't use his cell phone. The city was
> suddenly much more vulnerable to terrorism.[15]

THE ECONOMIC COSTS OF BLACKOUTS

In addition to the human misery, which was widespread,
the economic losses caused by the blackout were
enormous. The Anderson Economic Group estimated
the total cost of the blackout to be about $6.4 billion,
with most of the losses ($4.2 billion) the result of
lost income to workers and investors. Various levels
of government also incurred losses; estimates range

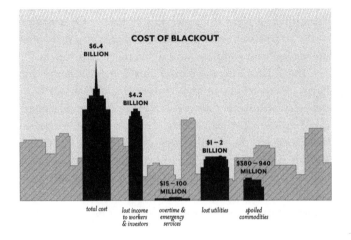

between \$15 and \$100 million in extra costs due to overtime and emergency service provision. There was another \$1 billion to \$2 billion in costs imposed on the affected utilities and between \$380 and \$940 million in costs associated with lost or spoiled commodities.[16]

In Ohio alone, the Ohio Manufacturers' Association estimated the direct costs of the blackout on Ohio manufacturers at more than one billion dollars. The organization explained: "Some 12,300 manufacturing companies in the state (representing approximately 55% of the manufacturers in Ohio were impacted with an average estimated direct cost of nearly \$88,000 each. All companies reporting indicated that the blackout caused a 'complete shutdown in operations.' The average duration of a plant shutdown was 36 hours. Over a third of the companies reported that the outage also disrupted deliveries from suppliers and deliveries to customers."[17]

But it isn't only huge blackouts that cause economic harm. Researchers at the University of California at Berkeley point out that even smaller, shorter blackouts can impose significant costs. In a study on electrical blackouts in the United States, Kristina Hamachi-LaCommare and Joe Eto estimate that blackouts could cost up to $130 billion every year and that about two-thirds of that (67 percent) is caused by short-term interruptions lasting less than five minutes.[18]

THE CAPACITY FACTOR

Reliability is an important factor in today's discussions about how we wish to develop our energy systems, because different types of power generation have different levels of reliability. And to understand this, we have to look not only at how much energy a given type of generator could produce but how often that generator actually runs at its peak level of output. This is called the "capacity factor" of energy production, and it's the ratio of how much energy is produced over a given period of time compared to how much could be produced if the generator ran at peak in that same period. Part of this calculation is based on how reliable the technology is, in terms of whether there are extensive periods of downtime due to maintenance. Another part has to do with the cost of fuel inputs: if it's cheaper in one place to buy coal- or nuclear-generated

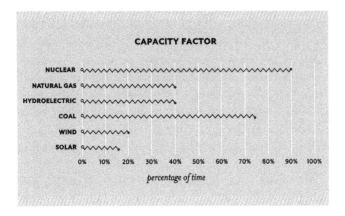

CAPACITY FACTOR

NUCLEAR

NATURAL GAS

HYDROELECTRIC

COAL

WIND

SOLAR

0% 10% 20% 30% 40% 50% 60% 70% 80% 90% 100%

percentage of time

electricity, utilities will do so, and a natural-gas plant might find itself idling, which cuts into its capacity factor.

Nuclear reactors, for example, have very high capacity factors, running at peak output 90 percent of the time. Combined-cycle natural-gas power plants have a capacity factor of about 40 percent, as does conventional hydroelectric power.[19] Coal-fired power plants have a capacity factor of about 74 percent, whereas wind power has a capacity factor of only 20 percent. That is, they are only producing their best possible output about one-fifth of the time.[20] Solar power is even worse, reaching capacity factors of only 19 percent in sunny Arizona and only 12–15 percent in often-cloudy Massachusetts.[21] Finally, capacity factors are important for another reason: the less reliable a given type of energy is, the more it has the ability to cause disruptions in the balance of energy flows in the grid, risking the kind of blackouts like the one discussed above. As Vaclav Smil, one of the leading scholars in the field of energy, observes, energy sources such as wind are useful, but only in small quantities: "Many studies have demonstrated that these variations cause no unmanageable problems, even in an isolated electricity generating system, as long as the total power installed in wind turbines is no more than about 10 percent of the system's overall output."[22]

Americans are generally fortunate in having highly reliable energy. When we flip the switch, the power flows. When we open the gas tap, the gas flows. But such reliability is not something that can be taken granted, as the people of the East Coast learned to their dismay in 2003.

Of course, as the saying goes, "There's no such thing as a free lunch," and that's true with energy use as well. Although energy use is vitally important to human welfare, there's no question that energy production, distribution, and use can cause harm to wildlife, ecosystems, and human health. In the next section, we'll discuss the relationship between energy use and the environment.

STUDY QUESTIONS

1. Various politicians make aggressive claims about powering the future with wind and solar energy. Discuss (in approximately 500 words) how known capacity factors impact such claims.

2. Aside from the problems mentioned above that occurred in the 2003 blackout, can you think of six other ways that sudden losses of power might cause harm?

3. Imagine you live in an apartment on the fifteenth floor of an apartment building. List the things that would be unavailable to you in the event that the electricity went off for a week.

4

ENERGY AND THE ENVIRONMENT

[The] dirtiest water and air are
not found in the rich countries,
rather they are found in the
developing nations. As pollution
is rapidly becoming a global
issue, worldwide prosperity
should be viewed as the
solution to, not the cause of the
problem.

*—Hans-Joachim Ziock, Klaus
Lackner, and Douglas Harrison*[23]

Most people know that energy production causes considerable environmental damage. And indeed, energy production, distribution, and use are responsible for much of the damage that humanity inflicts on the environment. Globally, energy production produces prodigious amounts of air pollution, water pollution, habitat destruction, landscape destruction, wildlife mortality, and much more.

But what people are often confused about is the nature of the relationship between energy use and environmental damage over time. Since the time of Paul Erlich and the *Population Bomb*, not to mention Al Gore's *Earth in the Balance*, environmental activists have asserted that there is a linear relationship between energy and the environment and that it's a bad one. In their equation, more humanity, plus more energy use, automatically equates to more damage. As Paul Erlich once famously opined about the possibility for unlimited energy for humanity, it would be "like giving a machine gun to an idiot child." Jeremy Rifkin, another environmental activist, said that it "would be the worst thing that could happen to our planet."[24]

THE ENVIRONMENTAL KUZNETS CURVE

But the above view of the relationship between energy and the environment is both naïve and misleading. Economists have long observed that there is a better way to look at the triad relationship of humans, energy, and the environment, and that is a much more optimistic

FIGURE 1. THE ENVIRONMENTAL KUZNETS CURVE

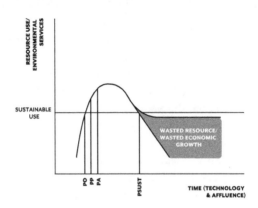

one, based on observations of how energy's impact on the environment changes as countries go through development. Rather than displaying a linear relationship between energy use and environmental degradation, the real relationship looks more like an inverted letter U. This relationship is generally called the environmental Kuznets curve, or the environmental transition curve.

Figure 1 is a graphic representation of this Kuznets curve. The bottom axis is economic growth, and the upright axis represents environmental use of a natural resource such as timber, water, or soil. The upright axis might also represent the use of environmental services such as diluting waste products in the air or the service one gets from a river's ability to break down a certain quantity of waste in a manner that harms neither fish nor people.

As the figure illustrates, for any given environmental resource, society passes through a series of phases. As countries develop, they use natural resources and environmental waste management services to build wealth with which the people satisfy their basic needs for housing, food, education, health care, mobility, and so on. If a country grows large enough, a society will often use more than its local environment can sustain. That is the point marked on the figure as Po, the point where overutilization of a resource commences. The horizontal line represents the sustainable-use level of the resource, which should be understood as a dynamic capacity that changes over time, as populations change and as climates fluctuate. It must be evaluated on an ongoing basis. The point of perception, Pp, where people notice they are overutilizing a resource, quickly follows, and people take steps to reduce their overuse, both as individuals and as a society. This is the point of action, or Pa. Finally, and usually in relatively short order, the overuse ends, and resource use is reduced, one hopes, to the maximum sustainable level, which is indicated in the figure as Psust. Driving environmental resource use to zero, whether it's the consumption of a given fuel or the use of nature's waste remediation ability, represents a massive amount of lost economic value that could be used for a myriad of good uses, such as alleviating poverty here and abroad or paying for the many entitlements that the United States has adopted over the last century.

AIR POLLUTION AND THE KUZNETS CURVE

For an example of the Kuznets curve, let's take the example of air pollution—the production and use of fossil fuels is a major source of air pollutants around the world. From mining to refining, virtually every step in energy production, conversion, distribution, and, often, use result in the emission of a variety of air pollutants, including coarse and fine particulate matter, oxides of nitrogen, sulfur oxides, ozone precursors, and yes, some of the greenhouse gases as well, including carbon dioxide, nitrous oxide, and methane. Air pollution takes a significant toll on human health, causing respiratory and cardiovascular problems in sensitive members of the population. And in the early days of our development, we certainly did foul our air. Who hasn't heard of the infamous smog of Los Angeles in the 1950s?

But here in the United States, we are long past the point of perception, which began around 1900. And the cleanup began shortly thereafter. As researcher Indur M. Goklany, a student of environmental transitions, points out, "By 1912 the federal Bureau of Mines reported that 23 of the 28 cities that had populations in excess of 200,000 were making some effort to control smoke."[25] And air pollution levels continue to decline sharply as newer technologies and pollution control devices combine to make our system of energy production

TABLE 3. AIR POLLUTION TRENDS, 1980 VERSUS 2008 AND 1990 VERSUS 2008

	1980 VS. 2008	1990 VS. 2008
Carbon Monoxide (CO)	-79	-68
Ozone (O3) (8-hr)	-25	-14
Lead (Pb)	-92	-78
Nitrogen Dioxide (NO2)	-46	-35
PM10 (24-hr)	NA	-31
PM2.5 (annual)	NA	-19
PM2.5 (24-hr)	NA	-20
Sulfur Dioxide (SO2)	-71	-59

Notes: Negative numbers indicate improvements in air quality; NA = Trend data not available; PM2.5 air quality based on data since 2000.

Source: United States Environmental Protection Agency, "Air Quality Trends," available at http://www.epa.gov/airtrends/aqtrends.html. Last updated October 8, 2009.

cleaner every year. Table 3 shows how air pollution levels here in the United States have improved, even as our energy use continues to increase.

But the situation is far worse in the developing world, where outdoor pollution takes a high toll, and indoor pollution is higher still. According to the World Health Organization:

> More than half of the world's population relies on dung, wood, crop waste or coal to meet their most basic energy needs. Cooking and heating

with such solid fuels on open fires or stoves without chimneys leads to indoor air pollution. This indoor smoke contains a range of health-damaging pollutants including small soot or dust particles that are able to penetrate deep into the lungs. In poorly ventilated dwellings, indoor smoke can exceed acceptable levels for small particles in outdoor air 100-fold. Exposure is particularly high among women and children, who spend the most time near the domestic hearth. Every year, indoor air pollution is responsible for the death of 1.6 million people— that's one death every 20 seconds.[26]

Furthermore, according to WHO:

indoor air pollution [is] the 8th most important risk factor and [is] responsible for 2.7% of the global burden of disease. Globally, indoor air pollution from solid fuel use is responsible for 1.6 million deaths due to pneumonia, chronic respiratory disease and lung cancer, with the overall disease burden (in Disability-Adjusted Life Years or DALYs, a measure combining years of life lost due to disability and death) exceeding the burden from outdoor air pollution five fold. In high-mortality developing countries, indoor smoke is responsible for an estimated 3.7% of the overall disease burden, making it the most lethal killer after malnutrition, unsafe sex and lack of safe water and sanitation.

THE IMPORTANCE OF UNDERSTANDING THE KUZNETS CURVE

Energy is clearly not environmentally benign—our use of energy causes air pollution, water pollution, land degradation, ocean degradation, and more. However, understanding the environmental transition curve suggests that as societies continue to develop, that environmental impact will reduce over time. Indeed, the environmental transition curve suggests that the single best thing we could do to minimize energy's impact on the environment is to not only maximize our own economic growth but also help developing countries to increase theirs, letting them switch to ever cleaner, less polluting forms of energy.

Caveats apply, of course—some economists argue that the environmental transition curve does not apply to all pollutants and all societies and that while it might work for local-area pollutants and resource protection, it may not work for global pollutants, such as soot or other greenhouse gases. They think that some rich countries might bring pollution to other parts of the world, as various businesses are forced to relocate to remain competitive. That may well be true, but it does not negate the idea of an environmental transition; it simply lengthens the time it takes to turn things around for certain global pollutants, because remediation then becomes dependent on other countries passing through their own environmental transitions.

THE DEEPWATER HORIZON OIL SPILL

On April 20, 2010, the Deepwater Horizon, a mobile, semisubmersible deep-sea oil-drilling rig leased by British Petroleum (BP), was completing a newly drilled well forty-one miles off the Louisiana coastline in the Gulf of Mexico when it exploded and sank, killing eleven oil-rig workers, injuring seventeen, and triggering the largest offshore oil spill in history. The spill took more than three months to stop and poured an enormous amount of oil into the Gulf of Mexico. Although estimates vary, the U.S. government estimates that the Deepwater Horizon spill released four million barrels of light sweet crude oil into the waters of the Gulf of Mexico.[27]

The oil released into the Gulf posed an ecologic threat to the coastlines of Mississippi, Alabama, Florida, Louisiana, and potentially points farther north, and between direct impacts of oil contamination and the indirect effects of fishing bans and negative side effects of remediation efforts, the Gulf coast sustained massive economic losses.

Nobody should deny that the Gulf oil spill was a serious environmental threat that merits serious study and strenuous efforts to avoid a similar incident in the future. At the same time, one must consider the realities we face as a country that uses vast quantities of oil: most of our transportation systems and the very base of our chemistry production rest on the use of oil and natural gas. If we are not to produce our own oil and natural gas, then we're going to have to get it from elsewhere, and therein lies the rub, for bringing in oil by tankers poses a more constant, predictable threat than does drilling for it and bringing it ashore by pipeline.

As my colleague Steven F. Hayward and I observed in a study published shortly after the start of the Deepwater Horizon spill, context is important.[28] Prior to the Deepwater Horizon spill, the difference between leaks at oil rigs and tankers was stark: "Over the last sixty years, there have been ten offshore-drilling accidents that released more than 5,000 tons of oil into ocean waters. During this same period, there have been seventy-two oil spills from tanker accidents that released 5,000 tons of oil or more—usually a lot more. In other words, for every offshore-drilling accident, there are seven major tanker spills and numerous tanker accidents of smaller size."[29]

FIGURE 2. TANKER OIL SPILLS VERSUS OFFSHORE-DRILLING SPILLS, 1957–2010

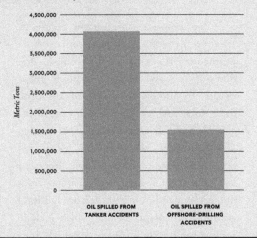

Figure 2 puts this in perspective.

We also observed that ecosystems are highly resilient, a fact that has been borne out in the Gulf of Mexico, where bacteria have consumed the vast majority of oil spilled out at sea, preventing harm to coastal ecosystems. And although some seem to find this surprising, they shouldn't have: the Gulf of Mexico is an ecosystem long adapted to oil leaks. As we observed:

> The amount of oil-based products entering the water each year from offshore production pales in comparison to the amount released through natural seeps or due to human consumption, disposal, and leakage of petroleum products. The NAS 2003 report brief *Oil in the Sea III* notes that "releases from extraction and transportation of petroleum represent less than 10 percent of inputs from human activity. Chronic releases during consumption of petroleum, which include urban runoff, polluted rivers, and discharges from commercial and recreational marine vessels, contribute up to 85 percent of the anthropogenic load to North American waters." Some estimate that the amount of oil-based products Americans pour down their household drains exceeds 300 million gallons (or about 1 million tons—much more oil than the Deepwater Horizon's upper spill estimate) each year. The NAS report brief estimates that natural oil seepage into the

northern Gulf of Mexico (the area closest
to the U.S. coastline) ranged from a low of
4,000 tons per year to as much as 17,000
tons per year; for the entire Gulf of Mexico,
the range is estimated to be 80,000 to
200,000 tons per year.[30]

There is no question that we should do all we
can to minimize the environmental impacts of oil
exploration and production. Both the ecosystem and
economic health of coastal environments require that
such care be taken, However, as we observed, there
are dangers in making hasty decisions. Prematurely
shutting down oil and gas development in the Gulf has
caused, and continues to cause, massive economic
losses and potentially the loss of a major industry that
underpins the economies of most Gulf states. And
turning to gasoline alternatives such as corn ethanol
poses its own massive threat to ecosystems, a threat
we will discuss in detail later in this book.

The first four chapters of this book focused on the
fundamental relationship of human civilization to energy
availability, affordability, reliability, and environmental
impacts. The next chapters will focus on questions

relating to proposed changes in our energy systems, changes based on concerns such as national security and renewable energy. But before we get to those topics, we have to start with a discussion about the nature of energy system transitions.

STUDY QUESTIONS

1. Psychologist Abraham Maslow defined humanity's "hierarchy of needs," which begin with physiological needs, then progress to safety needs, emotional needs, esteem, and finally self-actualization. In approximately 750 words, explain where you think environmental protection falls in that spectrum of needs.

2. This chapter shows how the environmental Kuznets curve applies to air pollution. In an essay of approximately 500 words, can you explain how the Kuznets curve would apply not only to pollution but to the use and overuse of other natural resources, such as coastal fisheries?

3. In approximately 500 words, can you explain why it would be undesirable to drive pollution or resource use down to zero, rather than trying to stabilize both pollution and resource use at a sustainable level?

5

ENERGY SYSTEM INERTIA AND MOMENTUM

I am tonight setting a clear goal for the energy policy of the United States. Beginning this moment, this Nation will never use more foreign oil than we did in 1977—never.

cont.

From now on, every new addition to our demand for energy will be met from our own production and our own conservation. The generation-long growth in our dependence on foreign oil will be stopped dead in its tracks right now and then reversed as we move through the 1980's, for I am tonight setting the further goal of cutting our dependence on foreign oil by one-half by the end of the next decade—a saving of over 4 1/2 million barrels of imported oil per day.

—Jimmy Carter[31]

Politicians are fond of promising rapid energy transitions. Whether it is a transition from imported to domestic oil or from coal-powered electricity production to natural-gas power plants, politicians love to make grandiose claims. Unfortunately for them (and often the taxpayers), our energy systems are a bit like an aircraft carrier: they're unbelievably expensive, they're built to last for a very long time, they have a huge amount of inertia (meaning it takes a lot of energy to set them moving), and they have a lot of momentum once they're set in motion. No matter how hard you try, you can't turn something that large on a dime, or even a few thousand dimes.

INERTIA AND MOMENTUM: AN OVERVIEW

In physics, moving objects have two characteristics relevant to understanding the dynamics of energy systems. Those two characteristics are *inertia* and *momentum*. Inertia is the resistance of objects to efforts to change their state of motion. If you try to push a boulder, it pushes you back. Once you've started the boulder rolling, it develops *momentum*, which is defined by its mass and velocity. Momentum is said to be "conserved," that is, once you build it up, it has to go somewhere. So a heavy object, like a football player moving at speed, has a lot of momentum, that is, once he's moving, it's hard to change his state of motion. If you want to change his trajectory, you have only a few choices: you can intercept him, transferring (possibly painfully) some of his kinetic energy to your own body, or you can approach alongside and slowly apply pressure to gradually alter his trajectory.

But there are other kinds of momentum as well. After all, we don't speak only of objects or people as having momentum; we speak of entire systems having momentum. Whether it's a sports team or a presidential campaign, everybody relishes having the "big mo," because it makes them harder to stop or deflect.

TECHNOLOGICAL MOMENTUM

One kind of momentum is *technological momentum*. When a technology is deployed, its impacts reach far beyond itself. Consider the incandescent bulb, a current *objet d'hatred* of many environmentalists and energy-efficiency advocates. The incandescent lightbulb, invented by Thomas Alva Edison, which came to be the symbol of inspiration, has been developed into hundreds, if not thousands, of forms. Today, a visit to a lighting store reveals a stunning array of choices. There are standard-shaped bulbs, oblong bulbs, flickering flame–shaped bulbs, colored globe–shaped bulbs, outdoor spotlights and floodlights, and more. It is quite easy, with all that choice, to change a lightbulb.

But the momentum of incandescent lighting doesn't stop there. All of those specialized bulbs led to the building of specialized light fixtures, from the desk lamp you study by, to the ugly but beloved hand-painted Chinese Foo Dog lamp you inherited from your grandmother, to the ceiling fixture in your closet, to the light in your oven or refrigerator, to the light that the dentist points at you, and, for the fortunate, to the antique chandelier in your family manor. It's easy to change a lightbulb, sure, but it becomes harder to change the bulb and its fixture.

And there's more to the story, because not only are the devices that house incandescent bulbs shaped to their underlying characteristics. Rooms and entire buildings have also been designed in accordance with how this one type of lighting reflects off of walls and windows. As lighting expert Howard M. Brandston points out:

> Generally, there are no bad light sources, only bad applications. There are some very laudable characteristics of the CFL [compact fluorescent lightbulb], yet the selection of any light source remains inseparable from the luminaire that houses it, along with the space in which both are installed and lighting requirements that

need to be satisfied. In the pursuit of more useful lumens-per-watt metric, one must match the luminaire to the space being illuminated. The lamp, the fixture and the room: all three must work in concert and for the true benefits of end-users. If the CFL should be used for lighting a particular space, or an object within that space, the fixture must be designed to work with that lamp, and that fixture with the room. It is a symbiotic relationship. A CFL cannot be simply installed in an incandescent fixture and then expected to produce a visual appearance that is more than washed out, foggy and dingy. The whole fixture must be replaced—light source and luminaire—and this is never an inexpensive proposition.[32]

And Brandston knows a thing or two about lighting, being the man who illuminated the Statue of Liberty.

LABOR-POOL MOMENTUM

Another type of momentum we have to think about when planning for changes in our energy systems is *labor-pool momentum*. It's one thing to say that we're going to shift 30 percent of our electricity supply from, say, coal to nuclear power in, say, twenty years. But it's another thing to have a supply of trained talent that would let you carry out the promise. That's because engineers, designers, regulators, operators, and all of the other skilled people needed for the new energy industry are specialists who have to be trained first (or retrained, if they're the ones being laid off in some related industry), and education, like any other complicated endeavor, takes time. And not only do our prospective new energy workers have to be trained, they have to be trained in the right sequence. One needs the designers, and perhaps the regulators, before the builders and operators, and each cohort of workers in training has to know there is work waiting for them once they graduate. In some cases, colleges and universities might have to change their training programs, adding another layer of difficulty, given the prevalence of tenure in academia.

ECONOMIC MOMENTUM

By far the biggest type of momentum that comes into play when it comes to changing our energy systems is *economic momentum*. The major components of our energy system, such as fuel production and refining and electrical generation and distribution, are very costly installations that have very lengthy life spans and that have to operate for long periods of time before the costs of development have been recovered. When investors put up their money to build, say, a nuclear power plant, they expect to earn that money back over the planned life of the plant, which is typically between forty and sixty years.[33] Some coal power plants in the United States have operated for more than seventy years![34] The oldest continuously operated commercial hydroelectric plant in the United States is on

New York's Hudson River, and it went into commercial service in 1898.[35]

As Vaclav Smil points out, "All the forecasts, plans, and anticipations cited above have failed so miserably because their authors and promoters thought the transitions they hoped to implement would proceed unlike all previous energy transitions, and that their progress could be accelerated in an unprecedented manner."[36]

When you hear people speaking of making a rapid transition toward any type of energy, whether it's a switch from coal to nuclear power, or a switch from gasoline-powered cars to electric cars, or even switching from an incandescent to a fluorescent light, understanding energy system inertia and momentum can help you decide whether their plans are feasible.

And such changes are not without their own trade-offs and unintended consequences, as we'll see later. But first, we'll discuss one of the other driving motivations invoked to persuade us to make the kind of rapid energy transitions that so many previous politicians have promised. That driver is the idea of "Energy Independence."

STUDY QUESTIONS

1. In approximately 500 words, discuss how the concept of physical momentum translates into other kinds of momentum. Besides the types of energy system momentum discussed above, can you think of other types of momentum that could slow down a planned energy transition?

2. Even if one could rapidly and radically transform our energy systems, would it be a good idea? In approximately 500 words, discuss some of the benefits or liabilities of accelerated modification to our energy systems.

6

ENERGY INDEPENDENCE AND SECURITY

Energy independence has since become a prize bit of meaningful-sounding rhetoric that can be tossed out by candidates and political operatives eager to appeal to the broadest cross-section of voters. When the U.S. achieves energy independence... America will be a self-sufficient Valhalla,

cont.

with lots of well-paying manufacturing jobs that will come from producing new energy technologies. . . . When America arrives at the promised land of milk, honey, and super-cheap motor fuel, then U.S. soldiers will never again need visit the Persian Gulf, except, perhaps, on vacation.

—*Robert Bryce*[37]

Energy independence. It has a nice ring to it, doesn't it? If you think so, you're not alone, because energy independence has been the dream of American presidents for decades, and never more so than in the past few years, when the most recent oil price shock has been partly implicated in kicking off the great recession.

As energy expert Robert Bryce points out, "Every U.S. president since Richard Nixon has extolled the need for energy independence. In 1974 Nixon promised it could be achieved within six years. In 1975, Gerald Ford promised it in 10. In 1977 Jimmy Carter warned Americans that the world's supply of oil would begin running out within a decade or so, and that the energy crisis that was facing America was "'the moral equivalent of war.'"[38]

DEFINING "INDEPENDENCE"

"Energy independence" and its rhetorical companion, "energy security," however, are slippery concepts that are rarely thought through. What is it we want independence from, exactly?

Most people would probably say that they want to be independent from imported oil. And indeed, the United States imports quite a bit of oil. According to the U.S. Energy Information Agency, we import about eleven million barrels of crude oil a day. But there are reasons that we buy all that oil from elsewhere.

INDEPENDENCE AND TRADE-OFFS

The first reason is that we need it to keep our economy running. There is no meaningful substitute for petroleum products used in transportation. As Bryce observes, "Regardless of the ongoing fears about oil shortages, global warming, conflict in the Persian Gulf, or terrorism, the plain, unavoidable truth is that the U.S., along with nearly every other country on the planet, depends on fossil fuels. And that dependence will continue for the foreseeable future."[39] Yes, there is a trickle of biofuel available, and more may become available, but as you'll read later on, most biofuels are a Faustian bargain, causing economic waste and environmental destruction. Yes, we can conserve fuel better, drive less, drive smaller cars, and live closer to work, but all of these options have costs along with benefits. There is no free lunch.

Second, Americans have basically decided that they don't really want to produce their own oil. For about forty years now, the American public, working through their politicians, have decided that they value the environmental quality they preserve over their oil imports from abroad. Vast areas of the United States—both on and offshore—have been put off-limits to oil exploration and production in the name of environmental protection. We could certainly reduce our petroleum imports if we wanted to by drilling offshore, drilling in the Arctic, and producing oil from the oil-rich shale formations that are plentiful here, but at what cost? To what extent are Americans really willing to endure the environmental impacts of domestic energy production in order to curtail imports?

Third, there are benefits to trade, and we trade in fossil fuels with about ninety different countries, including some that are our largest trading partners and nearest neighbors. Trade allows for economic efficiency, and when we buy things from places that have lower production costs than we do, we benefit. In fact, our leading source of imported crude oil is Canada, which sells us eighty-two million barrels of crude oil per month. Mexico sells us about thirty-six million barrels per month, just slightly less than Saudi Arabia. And although you don't read about this much, the United States is also a large *exporter* of oil products, selling about two million barrels of petroleum products per day to about ninety countries. Again, the top two customers for our exports are Canada and Mexico. Tables 4 and 5 show our top fifteen sources of imported oil, as well as the top fifteen destinations for our oil and petroleum product exports.

TABLE 4. TOTAL CRUDE OIL AND PETROLEUM PRODUCT IMPORTS BY COUNTRY OF ORIGIN, JUNE 2010

COUNTRY	THOUSANDS OF BARRELS IN JUNE 2010
Canada	81,978
Saudi Arabia	40,576
Mexico	36,231
Nigeria	33,284
Venezuela	26,955
Russia	22,791
Iraq	18,914
Algeria	16,496
Angola	12,755
Colombia	12,201
Brazil	9,247
UK	8,075
Virgin Islands (U.S.)	7,330
Kuwait	5,505
Ecuador	6,332

Source: U.S. Energy Information Administration, "U.S. Imports by Country of Origin," 2010, available at http://tonto.eia.doe.gov/dnav/pet/pet_move_impcus_a2_nus_ep00_im0_mbblpd_m.htm.

There is no question that the United States imports a great deal of energy and, in fact, relies on that steady flow to maintain our economy. When that flow is interrupted, we feel the pain in short supplies and higher prices. It is also true that some of our money does go to those who wish to do us harm and props up governments that we may well find heinous. At the same time, we derive massive

TABLE 5. TOTAL CRUDE OIL AND PETROLEUM PRODUCT EXPORTS BY COUNTRY OF ORIGIN, JUNE 2010

COUNTRY	THOUSANDS OF BARRELS IN JUNE 2010
Mexico	12,963
Canada	5,861
Netherlands	5,844
Brazil	4,468
Chile	3,252
Japan	2,890
Ecuador	2,717
Panama	2,417
China	2,292
Singapore	1,746
Bahamas Islands	1,744
Netherlands/Antilles	1,512
Italy	1,294
Columbia	1,270
Spain	1,220

Source: U.S. Energy Information Administration, "U.S. Imports by Country of Origin," 2010, available at http://tonto.eia.doe.gov/dnav/pet/pet_move_expc_a_EP00_EEX_mbbl_m.htm.

economic benefits when we buy the most affordable energy on the world market and when we engage in energy trade around the world.

When considering things such as energy "security" or energy independence, one has to strive for clear thinking and consider all of the potential trade-offs and potential unintended consequences involved with our decisions.

Our penultimate discussion focuses particularly on that latter phrase, "the potential unintended consequences involved with our decisions." Of all energy decisions in the last thirty years, perhaps the best cautionary tale about unintended consequences involves ethanol, the chemical that puts the oomph in a cocktail, and that is now a part of the gasoline that drivers pump into their gas tanks every time they "fill 'er up."

STUDY QUESTIONS

1. The United States depends on a variety of imports, including clothing, automobiles, consumer products, food, and more. In approximately 750 words, explain why politicians do not try to make the United States clothing independent, food independent, or car independent.

2. Robert Bryce argues that a better goal than independence is expanded interdependence so that countries we trade with have the best incentives to ensure our supply and face the competition to keep their prices low. In approximately 500 words, explain how trade in energy, rather than rendering us less secure, could make us more secure.

7

THE DANGER OF UNINTENDED CONSEQUENCES: THE ETHANOL FIASCO

The huge corn ethanol mandates imposed by Congress a few years ago may be the single most misguided agricultural program in modern American history.

cont.

That's saying something, but consider the program's impact: higher global food prices, increased air pollution from burning ethanol-spiked fuels, spreading dead zones in the Gulf of Mexico from a surge of fertilizer use, and strong evidence that growing a gallon of corn ethanol produces just as many greenhouse gases as burning a gallon of gas.

—*Robert Bryce*[40]

Another important factor to consider when pondering energy policy is the potential for unintended consequences emerging from governmental interventions in energy markets.

GOVERNMENT AND UNINTENDED CONSEQUENCES

All actions, private or governmental, can have unintended consequences, but governmental actions are particularly prone to unintended consequences because, first, governments often act on scales much larger than those of any private entity. The system of rules and regulations that the government has erected penetrates all aspects of the economy, and it is generally impossible to predict in advance what all of the consequences will be with even minor interventions in the economy, much less interventions in anything as complex as our nation's energy markets. There is also a problem of incentives and responsibilities that lead government actions to often go astray. That is, decision makers in government are rarely, if ever, responsible when things go wrong. If you, as a private person, take an action such as swapping out your incandescent lightbulbs for compact fluorescent bulbs, and you don't like the resulting color, or you find that you have to heat your house more in winter because the CFLs don't put out heat as incandescent bulbs do, you're the one who has to live with the consequences. When a regulator decides to ban incandescent bulbs, the regulator may have to deal with the same consequences you do, but that regulator also inflicts those consequences on millions of people and is not legally responsible if you wind up with headaches from bad lighting or if your energy bills go up because you turn up your household thermostat to make up for the heat that you used to get from a nearby lamp.

THE CASE OF ETHANOL

One of the most illustrative cases of unintended consequences in recent U.S. history has been the adoption of policies favoring the use of ethanol made from corn as a motor fuel. Ethanol, the chemical that gives alcoholic beverages their punch, has been a critically important compound for a very long time. In fact, it has been used for eight thousand years or so—going back to the Paleolithic era—and some research suggests that Stone Age people consumed alcoholic beverages. As commentator George Will recently pointed out, some have even argued that ethanol, as a component of beer and wine, was a driver in helping humanity transcend its original hunter-gatherer lifestyle and begin living in denser population clusters.

Automobile designers have long recognized ethanol's potential as motor vehicle fuel. The first American internal combustion prototype made by Samuel Morey in 1826 ran on ethanol, and it remained the dominant automotive fuel until 1908, when a combination of rapidly dropping gasoline prices and rapidly increasing ethanol prices led the Ford Motor Company to introduce the Model T, an early "flex-fuel" vehicle that could use either gasoline or ethanol. Gasoline/ethanol blends were used until the 1920s in the United States and the 1930s in Europe but were finally replaced by gasoline with lead additives, which was discovered to stop engine knocking. Ethanol blends were essentially off the market by 1940.

Ethanol languished in niche uses for the next thirty years, until the 1973 oil embargo made ethanol look attractive again as a replacement for gasoline and rekindled an interest in building ethanol distilleries. Gas lines and shortages in the 1970s even had people thinking about fermenting local fruit and vegetable waste and surplus to stretch their supplies of gasoline. When the price of gasoline started dropping in the 1980s, interest in ethanol dropped with it, but ethanol fuel was given another look in 1989 when air pollution regulators mandated the use of fuel oxygenates (including ethanol) to reduce summertime pollution. This delighted the farm lobby, which was happy to take taxpayer subsidies to turn a crop surplus (usually a bad thing) into fuel supplements demanded by air quality laws. Later discovery of the groundwater-polluting effects of methyl tertiary butyl ether (MTBE)—the oxygenate that was ethanol's main competitor—led several states to switch from MTBE to ethanol as a fuel oxygenate.

Recent events have sent the fortunes of ethanol producers skyrocketing, as ethanol fuel has come back with a subsidy-powered roar. In a speech on April 25, 2006, President George W. Bush, known to abstain from drinking the stuff, was nearly rhapsodic about ethanol as fuel: "Ethanol is a versatile fuel. And the benefits are easy to recognize when you think about it.... Ethanol is good for our rural communities. It's good economic development for rural America.... Ethanol is good for the environment.... Ethanol's good for drivers.... Ethanol's good for the whole country."

And Bush was excited about our ability to make ethanol as well: "The ethanol industry is booming.... There are now ninety-seven ethanol refineries in our country, and nine of those are expanding, and thirty-five more are under construction.... But what's really interesting, there are new plants springing up in unexpected areas, like the Central Valley of California, or Arizona, or, of course, in the sugar fields of Hawaii."

Finally, Bush assured us, more ethanol is on the way: "I am committed to furthering technological research to find other ways, other sources for ethanol.

We're working on research—strong research to figure out cellulosic ethanol that can be made from wood chips or stalks or switchgrass. These materials are sometimes waste products that are just simply thrown away."

Before examining the implications of the use of ethanol as fuel, let us review some of the basics of ethanol: what it is and where it comes from.

The Basics of Ethanol. Ethanol is a chemical made when yeast breaks down sugar molecules in a process called fermentation. Ethanol has two carbon atoms, six hydrogen atoms, and one oxygen atom. There are various kinds of alcohol, but ethanol is the one found in beer, wine, and spirits. It is also combustible, which is well known to anyone who has set brandy on fire while cooking—think Cherries Jubilee.

Alcohols such as ethanol and its chemical cousins make effective fuels because they give off a lot of energy when burned, just as liquid fossil fuels do. The advantage of alcohol is that it is liquid at room temperature, which makes it easy to transport and handle.

Ethanol can be made in a variety of ways, but the usual way is still the most common: yeast is fed sugar molecules isolated from fruits or vegetables, and it produces ethanol as a by-product. The ethanol is then distilled from a dilute solution by controlled heating in order to drive off ethanol vapor. In the United States, the primary raw material for producing fuel ethanol is corn, while in other countries, such as Brazil, the primary source is sugar cane.

The Many Downsides of Ethanol. While ethanol promoters make it sound as if ethanol is the solution to all our energy woes—dependence on foreign oil, diminishing oil stocks, the environmental consequences of energy use, the decline of the family farm, and so on—a considerable amount of research has shown that ethanol has far more peril than it does promise.

Ethanol and Greenhouse Gas Emissions. Though ethanol is often pitched as a good solution to climate change because it simply recirculates carbon in the atmosphere, there is more than one kind of greenhouse gas to consider. Ethanol, blended with gasoline, actually turns out to increase the formation of potent greenhouse gases more than gasoline does by itself. As far back as 1997, the U.S. Government Accountability Office determined that the ethanol production process produces relatively more nitrous oxide and other potent greenhouse gases than does gasoline. In contrast, the greenhouse gases released during the conventional gasoline fuel cycle contain relatively more of the less potent type, namely, carbon dioxide.

In 2006, Paul Crutzen, a Nobel Prize–winning chemist, confirmed these findings. Crutzen and his coauthors found that when the extra N_2O emission from biofuel production is calculated in "CO_2-equivalent" global warming terms and compared with the quasi-cooling effect of "saving" emissions of fossil fuel–derived CO_2, the outcome is that the production of commonly used biofuels, such as biodiesel from rapeseed and bioethanol from corn (maize), depending on nitrogen fertilizer uptake efficiency by the plants, can contribute as much or more to global warming by N_2O emissions than cooling by fossil fuel savings.

In June 2007, two Colorado scientists, Jan F. Kreider, an engineering professor at the University of Colorado, and Peter S. Curtiss, a Boulder-based engineering consultant, determined that carbon dioxide emissions from corn-based ethanol are worse than those of conventional gasoline and diesel fuel. They concluded that carbon emissions in the life-cycle sense are about 50 percent higher for ethanols than for traditional fossil fuels; such fuels are not the answer to global warming—they make it worse.

In February 2008, researcher Timothy Searchinger and colleagues calculated that "corn-based" ethanol, instead of producing a 20 percent savings, nearly doubles greenhouse emissions over 30 years and increases greenhouse gases for 167 years. Biofuels from switchgrass, if grown on U.S. corn lands, increase emissions by 50 percent.

Ethanol and Air Pollution. Although the U.S. Environmental Protection Agency (EPA) claims a net decrease in greenhouse gas emissions from using ethanol, they recognize that ethanol use is a problem for conventional air pollutants. Ethanol use, according to the EPA, will increase the emission of chemicals that lead to the production of ozone, one of the nation's most challenging local air pollutants. At the same time, other vehicle emissions may increase as a result of greater renewable fuel use. "Nationwide, the EPA estimates an increase in total emissions of volatile organic compounds and nitrogen oxides (VOC + NOx) between 41,000 and 83,000 tons [due to increased use of ethanol]. . . . Areas that experience a substantial increase in ethanol may see an increase in VOC emissions between 4 and 5 percent and an increase in NOx emissions between 6 and 7 percent from gasoline powered vehicles and equipment."

Increases in pollutants have also been shown at the state and local level. In 2004 a study released by the California Air Resources Board indicated that gasoline containing ethanol caused VOC emissions to increase by 45 percent when compared to gasoline containing no oxygenates. And in mid-2006, California's South Coast Air Quality Management District determined that gasoline containing 5.7 percent ethanol may add as much as seventy tons of VOCs per day into the state's air. For a sense of scale, consider that an air quality regulator in the region around Los Angeles can become employee of the month by coming up with a way of reducing emissions by one-tenth of a ton per day.

More recently, Mark Z. Jacobson, a researcher at Stanford University, estimated that switching to a blend of 85 percent ethanol and 15 percent gasoline—relative to 100 percent gasoline—may increase ozone-related mortality, hospitalization, and asthma by about 9 percent in Los Angeles and 4 percent in the United States as a whole.

Ethanol and Fresh Water Consumption. What may surprise many people is how much fresh water it takes to produce ethanol. In December 2006, scientists at Sandia National Laboratory in New Mexico issued a report, "Energy Demands on Water Resources," explaining that virtually all forms of energy production consume a lot of water. Petroleum refining, for example, consumes one to two and a half gallons of water per gallon of refined product. Colorado scientists Kreider and Curtiss estimate that refining a gallon of corn ethanol today requires thirty-five gallons of water. But that is only the beginning. Kreider and Curtiss estimate that three times as much water is needed to grow the corn that yields a gallon of ethanol. That brings the tally to 140 gallons of water per gallon of corn ethanol produced. If their calculation is correct, the 5.4 million gallons of corn ethanol used in America in 2006 required the use of 760 million gallons of fresh water.

And things do not look much better for ethanol made from cellulose crops, such as switchgrass. Kreider and Curtiss estimate that switchgrass would require between 146 and 149 gallons of water per gallon of ethanol produced from cellulose, depending on the scale of production. Thus, meeting the Bush administration's target of thirty-five billion gallons of renewable and alternative fuels production in the United States by 2017 with cellulosic ethanol would require about five trillion gallons of water per year. That is a bit more than the average annual flow of the Colorado River, which the Southern Nevada Water Authority lists at fifteen million acre-feet, or a little under five trillion gallons.

Ethanol and Water Pollution. In "Water Implications of Biofuels Production in the United States," the National Academy of Sciences (NAS) points out that if the United States continues to expand corn-based ethanol production without new environmental protection policies, "the increase in harm to water quality could be considerable." Corn, according to the NAS, requires more fertilizers and pesticides than other food or biofuel crops. Pesticide contamination is highest in the Corn Belt, and nitrogen fertilizer runoff from corn already has the highest agricultural impact on the Mississippi River. In short, more corn raised for ethanol means more fertilizers, pesticides, and herbicides in waterways; more low-oxygen "dead zones" from fertilizer runoff; and more local shortages in water for drinking and irrigation. Fertilizer runoff does not just pollute local waters; it creates other far-reaching environmental problems. Each summer, the loading of nitrogen fertilizers from the Mississippi via the Corn Belt hits the Gulf of Mexico, creating a large dead zone—a region of oxygen-deprived waters unable to support sea life that extends for more than ten thousand square kilometers. The same phenomenon occurs in the Chesapeake Bay, in some summers affecting most of the waters in the mainstern bay. A recent study by researchers at the University of British Columbia shows that if the United States were to meet its proposed ethanol production goals—fifteen to thirty-six billion gallons of corn and cellulosic ethanol by 2022—nitrogen flows to the Gulf of Mexico would increase by 10 to 34 percent.

Ethanol and Land Consumption. In a February 2008 *Science* article, researchers calculated that producing fifteen billion gallons of corn ethanol to meet U.S. ethanol goals would require the diversion of corn from 12.8 million hectares of U.S. cropland and would, in turn, bring 10.8 million hectares of additional land into cultivation. Locations would include 2.8 million hectares in Brazil, 2.3 million in China and India, and 2.2 million here in the United States.

Projected corn ethanol production in 2016 would use 43 percent of the U.S. corn land harvested for grain in 2004 that would otherwise be primarily used to feed livestock, requiring big land-use changes to replace that grain or causing sharp price hikes due to scarcity of grain raised for human and livestock consumption. And this does not include infrastructure requirements. As a low-density feedstock, corn or switchgrass would be required in massive quantities to produce enough ethanol to slake the thirst of America's transportation fleets. Bringing that heavy, woody biomass to an ethanol processing plant would require an extensive transportation infrastructure, most likely both truck and train. And because ethanol cannot be run through pipelines—it is both corrosive and strongly attracts water molecules from soil outside the pipeline at seams or pipeline cracks, leading to dilution of the ethanol—it will have to be moved again by tanker truck to the blending stations, where it will be mixed with gasoline, or to fueling stations directly if we ever run vehicles on pure ethanol.

Ethanol and Energy Security. The fundamental thermodynamic limitations of biofuels will render them little more than niche sources, barring massive technological breakthroughs. These limitations undercut claims that biofuels and renewables will increase America's military or energy security.

The problem is one of time and energy density: while nature spent millions of years concentrating solar energy in the forms of peat, coal, oil, and natural gas, all of the biofuels rely on sunlight to grow crops. Because such energy is extremely diffuse, the scale of land consumption and the labor required to gather massive quantities of low-density fuel quickly leads to diminishing returns.

As Rockefeller University researcher Jesse Ausubel points out, it would take one thousand square miles of prime Iowa farmland to produce as much electricity from biomass as from a single nuclear power plant.

And cellulosic ethanol is not the solution. In

May 2006, John Deutch, a chemistry professor at the Massachusetts Institute of Technology, concluded that producing enough ethanol from switchgrass to displace one million barrels of oil per day would require that twenty-five million acres of land—about thirty-nine thousand square miles—be planted in switchgrass. That is an area about the size of Kentucky. He concluded that we can produce ethanol from cellulosic biomass sufficient to displace one to two million barrels of oil per day in the next couple of decades, but not much more than that. Even cellulosic ethanol cannot provide enough liquid fuel for us to be able to stop importing foreign oil and being dependent on foreign governments.

Biofuels cannot displace enough of the world oil market to cause economic hardship to enemy regimes (we actually have peaceful relations with most of our major suppliers), nor will renewables put a dent in the pocketbook of terrorists. As terrorism expert Colonel G. I. Wilson points out, "Most insurgencies are low-tech in nature. Terrorists don't need oil money. For terrorists, the money flow doesn't come from oil, it comes from drugs, crime, human trafficking and the weapons trade." That should be self-evident, as many terrorist groups exist in countries that do not have oil wealth. The Tamil Tigers have used terror in Sri Lanka, the Basque separatist group Euskadi Ta Askatasuna has used it in Spain, and the Irish Republican Army has used it in Ireland—and none of them needed petrodollars for their actions. Nor, for that matter, do the Taliban or the

Palestinians. Osama bin Laden's family wealth was made in the construction industry.

Ethanol and Food Prices. As noted energy expert Robert Bryce has observed, many studies have shown that corn ethanol has caused higher food prices at home and abroad, as well as shifts in food crops. As Bryce notes, "In May 2007, the Center for Agricultural and Rural Development at Iowa State University released a report saying the ethanol mandates have increased the food bill for every American by about $47 per year due to grain price increases for corn, soybeans, wheat, and others. The Iowa State researchers concluded that American consumers face a 'total cost of ethanol of about $14 billion.' And that figure does not include the cost of federal subsidies to corn growers or the $0.51 per gallon tax credit to ethanol producers."[41]

Furthermore, Bryce observes that

the U.S.D.A. [U.S. Department of Agriculture], the federal agency that has long been one of the corn ethanol sector's biggest boosters, admitted that corn ethanol is driving up food prices. That's somewhat remarkable given that the agency's leaders have consistently downplayed the link. Nevertheless, in July 2008, the department released a report called "Food Security Assessment, 2007," which states very clearly that the biofuels mandates are pushing up food prices. The first page of the report says...."the persistence of higher oil prices deepens global energy security concerns and heightens the incentives to expand production of other sources of energy including biofuels. The use of food crops for producing biofuels, growing demand for food in emerging Asian and Latin American countries, and unfavorable weather in some of the largest food-exporting countries in 2006–07 all contributed to growth in food prices in recent years."[42]

While that admission is noteworthy, the importance of the July 2008 report lies with its projections about the growing numbers of people around the world who are facing food insecurity. And although the USDA report does not correlate this increasing food insecurity with soaring ethanol production, the connections are abundantly clear: as the United States uses more corn to make motor fuel, there is less grain available on the market. That means higher prices. And that's a key factor for residents of poor countries, who generally spend a higher percentage of their income on food than their counterparts in the developed world. For instance, in the United States only about 6.5 percent of disposable income is spent on food. By contrast, in India, about 40 percent of personal disposable income is spent on food. In the Philippines, it's about 47.5 percent. In some sub-

Saharan African countries, consumers spend about 50 percent of the household budget on food. And according to the USDA, "In some of the poorest countries in the region such as Madagascar, Tanzania, Sierra Leone, and Zambia, this ratio is more than 60 percent." The July 2008 USDA report goes on to say that the number of people facing food insecurity jumped from 849 million in 2006 to 982 million in 2007. And those numbers are expected to continue rising. By 2017, the number of food-insecure people is expected to hit 1.2 billion. And, says the USDA, "short-term shocks, natural as well as economic," could make the problem even worse.

STUDY QUESTIONS

1. Given all that is now known about the harmful nature of our country's corn ethanol policy, why do you think it remains as firmly entrenched as it does? Which interests benefit particularly from the ethanol mandates?

2. In approximately 750 words, explain the concept of unintended consequences. Can you suggest ways to minimize the possibility of unintended consequences of energy regulation?

Note: This chapter is derived from Kenneth P. Green, "Ethanol and the Environment" [Washington, DC: American Enterprise Institute, 2008]. Full citation data is available in that document, which can be found online at http://www.aei.org/outlook/28396.

CONCLUSION

We have surveyed some critical concepts needed to understand energy, its role in our society, its impacts on our lifestyle, its impact on the environment, and the importance of affordability and reliability. We have also discussed the meaning of energy independence, and we have looked at the potential for unintended consequences of even the best-intended interventions into our vast energy systems.

It is not my desire to tell anyone how to think about energy. Nobody can reasonably do that for another person. Rather, my hope is that this book will inspire readers to study energy more deeply and to question facile claims of simple solutions to complex energy-related challenges, regardless of who makes them. And most important, I hope this book will help readers formulate their own set of questions, think through their own values as they apply them to questions of energy policy, and think through what they believe is the best energy policy that our country might adopt.

ONLINE RESOURCES

CHAPTER 1

Mark Derr, "Of Tubers, Fire, and Human Evolution," http://www.nytimes.com/2001/01/16/science/of-tubers-fire-and-human-evolution.html?pagewanted=2.

Science Daily, "Light My Fire: Cooking as a Key to Human Evolution," http://www.sciencedaily.com/releases/1999/08/990810064914.htm.

Wikipedia, "Control of Fire by Early Humans." http://en.wikipedia.org/wiki/Control_of_fire_by_early_humans.

CHAPTER 2

Kenneth P. Green and Aparna Mathur, "Indirect Energy and Your Wallet" (Washington, DC: American Enterprise Institute, 2008), http://www.aei.org/outlook/100017.

Kenneth P. Green and Aparna Mathur, "Measuring and Reducing Americans' Indirect Energy Use" (Washington, DC: American Enterprise Institute, 2008), http://www.aei.org/outlook/29020.

CHAPTER 3

Renewable Energy Research Laboratory, University of Massachusetts at Amherst, http://www.ceere.org/rerl/about_wind/RERL_Fact_Sheet_2a_Capacity_Factor.pdf.

United States Association of Energy Economics, http://dialogue.usaee.org/index.php?option=com_content&view=article&id=95&Itemid=113.

Wikipedia, "Capacity Factor and Renewable Energy," http://en.wikipedia.org/wiki/Capacity_factor#Capacity_factor_and_renewable_energy.

CHAPTER 4

Jesse H. Ausubel, "The Liberation of the Environment, *Daedalus* (Summer 1996), http://phe.rockefeller.edu/Daedalus/Liberation/.

John Tierney, "Use Energy, Get Rich, and Save the Planet," *New York Times*, April 20, 2009, http://www.nytimes.com/2009/04/21/science/earth/21tier.html?_r=1.

CHAPTER 5

New World Encyclopedia, "Hydroelectricity," http://www.newworldencyclopedia.org/entry/Hydroelectricity#Oldest_hydroelectric_power_stations.

Wikipedia, "Momentum," http://en.wikipedia.org/wiki/Momentum.

CHAPTER 6

Robert Bryce.com (www.robertbryce.com) has many articles about energy, including many on energy independence and security.

CHAPTER 7

Kenneth P. Green, "Ethanol and the Environment" (Washington, DC: American Enterprise Institute, 2008), http://www.aei.org/outlook/28396.

ENDNOTES

1 Daniel B. Botkin, "Energy and Civilization," available at http://www.danielbbotkin.com/2007/03/19/energy-and-civilization/.

2 Nathanial Gronewold, "One-Quarter of World's Population Lacks Electricity," *Scientific American*, November 24, 2009, available at http://www.scientificamerican.com/article.cfm?id=electricity-gap-developing-countries-energy-wood-charcoal.

3 Ibid.

4 Quoted in Richard Wrangham, *Catching Fire: How Cooking Made Us Human* (Philadelphia: Basic Books, 2009), 1.

5 Wrangham, *Catching Fire*.

6 Ibid., 102.

7 Frances D. Burton, *Fire: The Spark That Ignited Human Evolution* (Albuquerque: University of New Mexico Press, 2009).

8 U.S. Energy Information Administration, "2005 Residential Energy Consumption Survey—Detailed Tables," available at http://www.eia.doe.gov/emeu/recs/recs2005/c&e/detailed_tables2005c&e.html (from http://www.eia.doe.gov/emeu/recs/recs2005/c&e/detailed_tables2005c&e.html, with per-household calculations by author).

9 Mark Cooper, "The Impact of Rising Prices on Household Gasoline Expenditures," Consumer Federation of America, September 2005, available at http://www.consumerfed.org/elements/www.consumerfed.org/file/energy/CFA_REPORT_The_Impact_of_Rising_Prices_on_Household%20Gasoline_Expenditures.pdf.

10 Stephen Budiansky, "Math Lessons for Locavores," *New York Times*, August 19, 2010, available at http://www.nytimes.com/2010/08/20/opinion/20budiansky.html.

11 Hiroko Shimizu, "In Praise of the 10,000 Mile Diet," *PERC Report* 28, no. 1 (Spring 2010), available at http://www.perc.org/articles/article1225.php.

12 Pierre Desrochers and Hiroko Shimizu, "Yes, We Have No Bananas: A Critique of the 'Food Miles' Perspective," *Mercatus Policy Series Policy Primer* No. 8, 2010, available at http://mercatus.org/publication/yes-we-have-no-bananas-critique-food-miles-perspective.

13 Jaime Holguin, "Biggest Blackout in U.S. History," CBS News, August 15, 2003, available at http://www.cbsnews.com/stories/2003/08/15/national/main568422.shtml.

14 Daniel B. Botkin, *Powering the Future: A Scientist's Guide to Energy Independence* (Upper Saddle River, NJ: Pearson Education).

15 Ibid.

16 Patrick L. Anderson and Ilhan K. Geckil, "Northeast Blackout Likely to Reduce US Earnings by $6.4 Billion," AEG Working Paper 2003-2, August 19, 2003, available at http://www.andersoneconomicgroup.com/Portals/0/upload/Doc544.pdf.

17 Electricity Consumers Resource Council (ELCON), *The Economic Impact of the August 2003 Blackout* (Washington, DC: Electricity Consumers Resource Council, 2004), available at http://www.elcon.org/Documents/EconomicImpactsOfAugust2003Blackout.pdf.

18 Kristina Hamachi-LaCommare and Joseph H. Eto, "Understanding the Cost of Power Interruptions to US Consumers," Ernest Orlando Lawrence Berkeley National Laboratory, September 2004, available at http://certs.lbl.gov/pdf/55718.pdf.

19 United States Energy Information Administration, "Average Capacity Factors by Energy Source," 2010, available at http://www.eia.doe.gov/cneaf/electricity/epa/epat5p2.html.

20 Nicolas Bocard, "Capacity Factor of Wind Power: Realized Values vs. Estimates," 2008, available at http://www.hnsboroesc.outlierproductions.com/Resources/CapacityFactorOfWind.pdf.

21 Renewable Energy Research Laboratory, University of Massachusetts at Amherst, "Wind Power: Capacity Factor, Intermittency, and What Happens When the Wind Doesn't Blow?" available at http://www.ceere.org/rerl/about_wind/RERL_Fact_Sheet_2a_Capacity_Factor.pdf.

22 Vaclav Smil, *Energy Myths and Realities* (Washington, DC: AEI Press, 2010).

23 Hans-Joachim Ziock, Klaus Lackner, and Douglas Harrison, "Zero Emission Coal," *Energy 2000: The Beginning of a New Millennium* (Lancaster, PA: Technomic Publishing, 2000), 1274.

24 Paul Ciotti, "Fear of Fusion: What If It Works," *Los Angeles Times*, April 19, 1989.

25 Indur M. Goklany, *Clearing the Air: The Real Story of the War on Air Pollution* (Washington, DC: Cato Institute, 1999).

26 World Health Organization, "Indoor Air Pollution and Health," Fact Sheet No. 292, June 2005, available at http://www.who.int/mediacentre/factsheets/fs292/en/index.html.

27 Cutler J. Cleveland et al. "The Deepwater Horizon Oil Spill," *The Encyclopedia of the Earth*, available at http://www.eoearth.org/article/Deepwater_Horizon_oil_spill.

28 Kenneth P. Green and Steven F. Hayward, "The Dangers of Overreacting to the Deepwater Horizon Disaster," *AEI Energy and Environment Outlook*, June 2010, available at http://www.aei.org/outlook/100965.

29 Ibid.

30 Ibid.

31 Jimmy Carter, "Crisis of Confidence," July 15, 1979, available at http://www.cartercenter.org/news/editorials_speeches/crisis_of_confidence.html.

32 Howard M. Brandston et al., "Research into the Effects and Implications of Increased CFL Use," Professional Lighting Designers Association, March 2010, available at http://greenpages.pld-a.org/wp-content/uploads/2010/04/2010-03_Final-Report_comprehensive-email.pdf.

33 Nuclear Energy Agency, International Energy Agency, Organisation for Economic Co-operation and Development (OECD), *Projected Costs of Generating Electricity, 2005 Update* (Paris, France: International Energy Agency, 2005), 35.

34 Ibid.

35 *New World Encyclopedia*, "Hydroelectricity," available at http://www.newworldencyclopedia.org/entry/Hydroelectricity#Oldest_hydroelectric_power_stations.

36 Vaclav Smil, *Energy at the Crossroads* (Cambridge, MA: MIT Press, 2003).

37 Robert Bryce, "The Delusions of 'Energy Independence,'" *Innovations* 3, no. 4 (Fall 2008).

38 Carter, "Crisis of Confidence."

39 Bryce, "Delusions of 'Energy Independence.'"

40 Robert Bryce, "The Corn Ethanol Juggernaut," *environment360* (Yale University), http://e360.yale.edu/feature/the_corn_ethanol_juggernaut/2063/.

41 Reprinted with permission from http://www.counterpunch.org/bryce02052009.html.

42 Ibid.

Kenneth P. Green, a biologist and environmental scientist by training, has studied public policy involving risk, regulation, and the environment for more than sixteen years at public policy research institutions across North America. He is the author of numerous policy studies, magazine articles, newspaper columns, encyclopedia and book chapters, and a textbook for middle school students entitled *Global Warming: Understanding the Debate* (Enslow Publishers, 2002). Mr. Green has testified before regulatory and legislative bodies at both state and federal levels, and speaks frequently to the public and the media. He has twice served as an expert reviewer for the United Nations' Intergovernmental Panel on Climate Change.

Printed in the USA
CPSIA information can be obtained
at www.ICGtesting.com
JSHW082221140824
68134JS00015B/655

9 780844 772042